A Pocket Guide to
HAWAI'I'S
FLOWERS

Photography by Douglas Peebles
Text by Leland Miyano

MUTUAL PUBLISHING

A Pocket Guide to
HAWAI'I'S FLOWERS

HAWAI'I IS A LAND OF SUPERLATIVES. THE VOLCANIC ISLANDS ARE THE PEAKS OF SOME OF EARTH'S LARGEST MOUNTAINS. WE ARE THE MOST ISOLATED LAND MASS OF THE WORLD AT THE CENTER OF THE PACIFIC "RIM OF FIRE." VOLCANOES ARE STILL ACTIVELY ADDING NEW LAND TO OUR STATE. GROWING UPON THIS DYNAMIC FOUNDATION IS A RICH MANTLE OF VEGETATION. HAWAI'I IS A PARADISE WITH A WIDE VARIETY OF ECOLOGICAL ZONES RANGING FROM STRAND SEASHORE TO TROPICAL RAIN FOREST TO DRY ALPINE AND PRACTICALLY EVERY CONDITION IN BETWEEN. IT WAS IN THIS SETTING THAT AN ENDEMIC FLORA—UNLIKE ANY OTHER IN THE WORLD—EVOLVED.

The arrival of the Polynesians began the plant introductions that continue to this day. Hawai'i now has one of the largest tropical and subtropical assemblages from around the globe. This tapestry of native and exotic plants is what the observer will encounter today. Blessed with a mild climate, flowers in Hawai'i bloom throughout the year. Flowers are a plant's expression of love, and garlands of flowers called leis are traditionally given in this spirit. Hawai'i is full of aloha, literally, "in the presence of the breath of life." Certainly this is an appropriate description of a land blooming with a vibrance of life and activity. What better way is there to celebrate all that is the best in Hawai'i than the simple appreciation of flowers?

Design by Angela Wu-Ki

Fifteenth Printing, October 2015

ISBN-10: 1-56647-149-4
ISBN-13: 978-1-56647-149-7
Library of Congress Catalog
Number 96-79001

Mutual Publishing, LLC
1215 Center Street, Suite 210
Honolulu, Hawaii 96816
Telephone (808) 732-1709
Fax (808) 734-4094
e-mail: info@mutualpublishing.com
www.mutualpublishing.com

Printed in Taiwan

Table of Contents

Ginger Family

❁ The gingers are from a large family of tropical herbs primarily originating in the Indo-Malaysian region. Edible ginger root, tumeric, and cardamom are some of the spices derived from these plants. Fibers, dyes, and medicines are also found in this family of useful tropicals. As ornamentals, few can compare to the lush landscaping effect of flowering ginger plants. With about 700 species and many varieties, there is a great diversity of floral form from delicate white and yellow gingers to the robust inflorescences of the shell and red gingers.

RED GINGER

Alpinia purpurata
'Awapuhi-'ula'ula

The vivid red color and strong form of the red ginger contribute to its popularity as a landscape plant and as cut flowers. The inflorescences are borne terminally on the foliage stalks and bloom from spring to late fall. The true flowers are white and largely hidden between the showy red bracts, which form a club-like cluster. The blooms add long-lasting color and structure to many tropical floral arrangements.

SHELL GINGER

Alpinia zerumbet

'Awapuhi-'luheluhe

Shell ginger is a popular landscaping plant that is native to Burma and India. In Southeast Asia, the pith of the young shoots are eaten and the leaves are used for wrapping foodstuffs. The fibers of this plant formerly had commercial importance and may someday be used, as natural fibers regain popularity. Blooming in spring to late fall, the flower clusters resemble carefully strung strands of pink and white shells. The Hawaiian name, 'Awapuhi-'luheluhe, means "drooping ginger" and refers to the pendant habit of the inflorescences.

BLUE GINGER

Dichorisandra thysiflora

The superficial resemblance to the family Zingiberaceae has caused this Brazilian native spiderwort to be called the blue ginger. It has tall, cane-like stems topped with whorls of leaves and periodic inflorescences of blue-purple. The individual flowers are sometimes used for leis but they do stain fabric.

TUMERIC

Curcuma domestica

'Olena

Tumeric is famous as a component of curry seasonings and has been cultivated for thousands of years in the Indo-Malaysian region of its origin. It is also the source of yellow dyes and medicine. The large green and pink bracts of the inflorescence are very showy and appear in late summer or early fall. After blooming, the entire plant dies back and lies dormant until the spring brings a flush of new growth.

TORCH GINGER

Phaeomeria magnifica

'Awapuhi-ko'oko'o

Magnificent is only one way to describe the torch ginger. The plant grows to about 20 feet tall in large clumps. In late spring to early winter, 5-foot stems topped with fiery red inflorescences, resembling torches, appear. In Malaysia, the young floral shoots are used in the preparation of curries. The Hawaiian name, 'Awapuhi-ko'oko'o, means "walking stick ginger," and the strong cane-like stems are well suited for this use.

KAHILI GINGER

Hedychium gardnerianum

Kahili ginger is native to Nepal and Sikkim and was introduced to Hawai'i. When it blooms in late spring to fall, its flower clusters resemble the cylindrical, feathered *kahili* staffs of Hawaiian royalty. Certainly the *kahili* ginger bears its inflorescences proudly on stems up to 8 feet tall and lives up to its noble namesake.

INDONESIAN GINGER

Tapeinochilus ananassae

Indonesian ginger is actually a member of the Costaceae, which resemble gingers in both foliage form and inflorescences. The waxy bloom of this Moluccan native can be best described as bright red clubs that feel like hard plastic. It would be difficult to convince someone seeing and touching this inflorescence for the first time that it is not artificial. Yellow flowers are tucked among the hard red bracts, which are tipped with thorns to discourage predators. The foliage stalks grow with attractive spiral whorls of leaves, and the floral stalks appear separately in late spring to early fall.

Flowering Vines

❀ Vines give a fast-growing blanket of green and are often used as symbols of the tropics and the encroaching jungle. In bloom, the sheer mass of flowers borne by some vines is truly spectacular and contributes to their use in Hawai'i landscapes. Vines, with their serpentine twists, are often used effectively to soften hard architectural lines, or simply to emphasize wild abandon in the jungle garden.

PASSION FLOWER

Passiflora species

Passiflora is a genus with about 430 species known for its unusual flowers. Some species are prized for their edible fruits and Passiflora edulis has commercial value as a popular juice. All passion flowers have numerous seeds and several species are naturalized in Hawai'i. While the edible varieties provide hikers with a succulent treat, their vining habit and rapid dispersal make passion flowers a major noxious weed in Hawai'i's native forests.

RED JADE VINE

Mucuna bennettii

Red jade vine has long, cascading clusters of brilliant red flowers 3 to 4 feet long. It is rarer than the green jade vine, but it is very popular in gardens that can accommodate the vine. The red jade vine originated in New Guinea but flourishes in its adopted home.

GREEN JADE VINE

Strongylodon macrobotrys

Green jade vine flowers have a beautiful sea-green color that is almost beyond description. They bloom in pendant inflorescences up to 4-1/2 feet long and put on a spectacular show. This Philippine native is often used in leis, but the flowers will stain clothes if the lei is bruised.

CUP OF GOLD

Solandra hartwegii

Cup of gold is a large Mexican vine of the nightshade family that includes tomatoes, chili peppers, and tobacco. The short-lived flowers bloom in spring and are spectacular, fragrant, yellow goblets about 6 inches in diameter.

WOOD ROSE

Merremia tuberosa

Pilikai

The yellow-orange flowered wood rose is so named for its brown, hard, fruiting capsules and large petal-like sepals. These rose-like structures are popular in dried floral arrangements and will last for many years. The vine is a large, tropical American native with a tuberous root.

(right)

KUHIO VINE

Ipomea horsfalliae

The Kuhio or princes' vine is originally from the West Indies but popularly named after Prince Kuhio because he grew this plant on his Waikiki property. It bears large clusters of brilliant red flowers in fall to late winter and is a common vine for trellises and fences.

MORNING GLORY

Ipomea species

Koali or Kowali

The morning glory family, Convolvulaceae, has about 50 genera and 1,000 species, including Ipomea batatas, the sweet potato. In Hawai'i, the morning glory is found from the coast up to 2,000 feet in elevation. The attractive, bell-shaped flowers open in the morning as a bluish-violet color which changes gradually to pink. The plant is used medicinally as an external application for bruises. Morning glories play an important role in Polynesian mythologies and, in the Hawaiian myth of Kawelu and Hiku, the *koali* vine is used as a swing into the underworld.

GLORY BUSH

Tibochina urvilleana var. urvilleana

Tibochina is a genus of about 350 species of herbs, shrubs, or trees. In cultivation, the group is known for its beautiful flowers and easy care. Unfortunately, glory bushes are in the family Melastomaceae, which includes the most invasive and noxious weed species. Koster's curse, Clidemia hirta, forms an infestation in many of the native forests in Hawai'i.

BOUGAINVILLEA

Bougainvillea spectabilis, B. glabra, and cultivars

Bougainvilleas are widely cultivated in Hawai'i for their colorful inflorescences. The true flowers, small and tubular, are inserted amongst the showy bracts that range in color from white to orange and red to purple. Bougainvilleas are native to Brazil and are either large woody vines or trees. In cultivation they are often severely pruned into shape to confine their growth and keep their numerous thorns at bay.

Flowering Trees

❁ Blooming trees form an important protion of the floral inventory of Hawai'i. Trees range in size from large shrubs to massive giants and, when they flower, the show is often spectacular. After flowering, the spent blooms carpet the ground and provide a bit of "fall" color in the tropics. Fragrance is also a feature of many trees and, although some of the lightly perfumed blooms may escape detection, some trees advertise their nectar stores with heady scents that are hard to miss.

RAINBOW SHOWER TREE

Cassia javanica X fistula

These ever-popular trees are the result of a cross between a pink-and-white shower tree (C. javanica) with a golden shower tree (C. fistula). The hybrid trees produce copious flower clusters of peachy orange to orange and red during the months of March to August. The show of color is so magnificent that this tree leaves a lasting impression on those who witness the blooming. In Hawai'i the effect is compounded by the sheer number of trees which are used liberally along streets, parks, and private properties. The falling petals also form a wonderful display that carpets the surrounding area. Being a hybrid, this tree rarely forms the hanging seed pods that are characteristic of the parent plants.

GOLDEN TREE

Tabebuie donnell-smithii

The gold tree grows to 75 feet tall in its native Mexico and Central America. The flowers are like bells with five flaring, frilly lobes. The gold tree can bloom any time, but most often blooms in spring when the tree is bare of leaves.

UMBRELLA TREE

Schefflera actinophylla

Umbrella trees are native to Australia and commonly grown as small house plants. In the ground they grow rapidly to heights of 40 feet with spare branching and a flat canopy. The name umbrella tree is derived from the large palmately compound leaves that are radially arranged like the spokes of an umbrella. This tree is often called the octopus tree, due to the flower spikes that spread from a central point like the outstretched arms of a 6-foot octopus. To add to the effect, each spike is covered with pink flowers and red globose fruits that resemble suckers on a boiled, red octopus.

ROSE APPLE

Syzygium jambos

'Ohi'a loke

Rose apples are rarely cultivated in Hawai'i, but have naturalized in many areas and are commonly encountered on hiking trails. In India this tree is said to bear the golden fruits of immortality. The seeds produced gold and the juice formed the river Gambu, which had healing powers. Buddha is sometimes depicted under a rose apple tree. The tree grows to about 30 feet tall, with 3-inch, tufted yellowish blooms that appear in the spring. The crisp yellow fruit has a delicate flavor like the scent of roses.

ROYAL POINCIANA

Delonix regia

'Ohai-'ula

Endemic and endangered in its native Madagascar, the royal poinciana is widely cultivated and common in Hawai'i. The buttressed roots, gnarled branches, fern-like foliage, and flat canopy provide the mature tree with great character. However, it is the spectacular show of the flowering season that the tree is most famous for. The red-to-orange or yellow flowers appear in late winter on some trees, with other trees blooming in September. June and July are the peak months of blooming, with many trees in full flower and red carpets of spent blooms littering the ground.

CANDLENUT TREE

Aleurites moluccana

Kukui

Kukui is the official State tree of Hawai'i and forms a major component of the lower mountain forests where it is easily distinguished by its pale green foliage. The flowers are small, white blooms held in large clusters that are woven with the leaves into leis to represent the island of Molokai. Dyes, fuel, medicine, food, oil, ornament, gum, and wood are among the uses this tree held for the Native Hawaiian population. Today, many of these uses continue, and the visitor is often greeted with the presentation of a polished kukui nut lei upon arrival.

ROSE-FLOWERED JATROPHA

Jatropha hastata

This Cuban native grows as a slender shrub or small tree to 7 feet tall. The red flowers are arranged in clusters amidst the dark-green foliage. The trees are well suited to hot, dry areas and they suffer little effect from droughts that wilt many other trees.

(left)

PINK TECOMA TREE

Tabebuia pentaphylla

The pink tecoma is a tropical American tree species that is widely grown for its attractive foliage and abundance of flowers that bloom sporadically throughout the year. The tree is very tough once established and is used frequently for street planting.

AFRICAN TULIP TREE

Spathodea campanulata

The African tulip tree is often used in public areas where space can accommodate the buttressed trunks that rise up to 70 feet. The scarlet, cup-shaped flowers are borne in clusters and bloom through most of the year. There is also a beautiful, yellow-flowered form of this tree. Following the blooms are canoe-shaped capsules that rise stiffly in bunches.

CORAL TREE

Erythrina crista-galli

This Brazilian native is widely cultivated for its gnarled growth habit, medium size to 30 feet, and the beautiful maroon-red flowers which bloom from December to April. Although it makes beautiful leis, the coral tree blossoms are rarely used in this manner.

ERYTHRINA

Erythrina abyssinica

This African transplant is a small tree up to 20 feet tall, with an open crown and a strongly furrowed bark bearing many thorns. The brilliant red flowers are held in torch-like clusters and bloom from the bottom up. The flowering season is a long one, with the slowly opening inflorescences lasting from May to August. The flowers are followed by pods that resemble strands of wooden beads. The seeds are small, bright-red beans marked with a jet-black dot that are sometimes strung as leis.

PLUMERIA

Plumeria acuminata, P. obtusa, P. rubra

Pua melia

Plumeria is a tropical American native so common in Hawai'i that it is usually the first flower the visitor encounters in the form of leis that are presented upon arrival. The waxy, fragrant, and long-lasting spring flowers are borne in clusters on trees that look somewhat succulent. The plants are, in fact, quite drought resistant and are common sights in dry areas where other plants may suffer. The colors of the flowers range from white to yellow to red and maroon, and there are many named varieties. The trees are of easy culture and grow from cuttings. The plant exudes a poisonous, milky sap when injured, but this has not detracted from its popularity. Outside of Hawai'i the plumeria is often known as the frangipani.

PROTEAS

PROTEAS AND BANKSIAS ARE IN THE FAMILY PROTEACEAE, WHICH ALSO INCLUDES THE MACADAMIA NUT TREE. THESE NATIVES OF SOUTH AFRICA AND AUSTRALIA HAVE A WIDE RANGE OF FOLIAGE HABIT AND FLORAL FORM. THE GENUS PROTEA IS DERIVED FROM THE GREEK GOD PROTEUS, WHO COULD ASSUME MANY SHAPES. A PROPER DESCRIPTION CANNOT BE ENCAPSULATED IN A SHORT STATEMENT, ALTHOUGH "MAGNIFICENT," "FANTASTIC," AND "AMAZING" COME TO MIND. PROTEAS WERE INTRODUCED TO THE COOLER ELEVATIONS (2,000 TO 4,000 FEET) OF HAWAI'I IN 1965 AND NOW FORM THE BASIS OF A SUCCESSFUL CUT-FLOWER INDUSTRY. PROTEAS AND BANKSIAS HAVE COLLECTIONS OF FLOWERS TIGHTLY GROUPED IN INFLORESCENCES. PROTEAS ALSO HAVE MODIFIED LEAVES CALLED BRACTS THAT RESEMBLE PETALS, ADDING TO THE ILLUSION OF A SINGLE FLOWER. THESE INFLORESCENCES HAVE FORMS THAT ARE OFTEN CLUB-LIKE, WITH A RANGE OF TEXTURES THAT INCLUDE THE HAIRY AND FUZZY. THE BLOOMS ARE VERY LONG LASTING IN ARRANGEMENTS AND ARE USED EITHER FRESH OR DRIED. TO OFFSET THE WONDERFUL FLOWERS ARE EQUALLY INTERESTING FOLIAGE FORMS WHICH ARE ALSO USED IN ARTFUL ARRANGEMENTS.

KAHILI FLOWER

Grevellea banksii

Ha'iku

This small tree, to 20 feet tall, is an Australian native with red or cream-colored flowers borne in cylindrical spikes. Parts of this plant are poisonous and the hairy ovaries of the flowers may trigger an allergic reaction in some people. Because of this, *kahili* flowers are used only for head leis. The Hawaiian name, Ha'iku, is derived from the location on Maui where this plant was first introduced.

ORANGE FROST

Banksia prionotes

This is a fast-growing tree up to 30 feet high that bears terminal inflorescences on branches covered by long, leathery leaves that appear to be trimmed by pinking shears. The specific name is derived from the Greek prion (a saw), which accurately describes these leaf margins. The 6-inch, club-like inflorescences may bloom twice a year, which is a welcome feature for the cut-flower trade. This banksia is native to Western Australia but has adapted nicely to the cool, dry upland areas of Hawai'i.

KING PROTEA

Protea cynaroides

The king protea is well named, as its flowerhead can be one foot across. The combination of large size, felty texture, and pastel colors makes for a most improbable sight to the uninitiated viewer. This protea is the national flower of South Africa and has adapted well to the dry, cool altitudes of Hawai'i. Having several varietal forms with different blooming seasons insures flowering through most of the year.

(inset)

RASPBERRY FROST BANKSIA

Banksia menziesii

The Latin species name of this banksia honors Archibald Menzies, the naturalist-surgeon on the Discovery expedition of 1791-1795. The plant itself was discovered in 1827 in Western Australia. In growth habit there are both tree and shrub forms with leaves that are dull green above and silvery pale underneath. The inflorescences are borne terminally on the branches and form colorful clubs on the plants. The trees can be up to 30 feet tall and are adapted to resist fire damage.

PINK MINK PROTEA

Protea neriifolia

This protea can be found in colors ranging from white to yellow and green, but it is best known in its pink phase. This was the first South African plant to be described over 350 years ago. The club-like inflorescences, appearing in August to March, are usually seen as buds or partially open blooms. The common name comes from the pink bracts which have a dark fringe that looks like mink fur.

PINCUSHIONS

Leucospermum Species

Pincushion proteas resemble varicolored sea urchins with curved spines, or the burst of fireworks on New Year's Eve. There are several species and cultivars of pincushion that range in color from pink to orange and gold. These inflorescences are the only proteas used commonly in leis due to their smaller size. Properly cared for, these leis can last for a month. This South African plant is adapted to the cool and dry higher elevations of Hawai'i, where one bush can produce over 1,000 flowerheads in a season.

Hibiscus

✽ The hibiscus, or aloalo, is the State flower of Hawai'i and its showy blossoms are a common feature of local landscapes. Hibiscus belongs in the family Malvaceae, which includes cotton and 'ilima (Sidax fallax), the official flower of O'ahu. There are a considerable amount of hibiscus species and cultivars, but all are transient blooms lasting only a day. The flowers may be white, yellow, orange, and red, or combinations of colors. New hybrids change the shade and the forms of the flowers and some are double blooms or fringed—the possibilities are endless. Besides its floral beauty, hibiscus is used for food (okra), tea, fibers, wood, dyes, perfumes, and medicines.

CORAL HIBISCUS

Hibiscus schizopetalus

Aloalo Koʻakoʻa

This East African native shrub grows with gracefully arching branches to 12 feet tall. The pendant, red flowers have highly fringed petals that bend upwards and back, with the staminal column hanging down. The relatively small size of the blooms compared to the mass of the plant has not diminished the popularity of this shrub in landscaping. The flowers are also used to create new hybrids to add to the large variety of forms now grown.

NATIVE WHITE HIBISCUS

Hibiscus arnottianus

Kokia keokeo

This endemic hibiscus is a tall shrub or tree to 30 feet high—growing at altitudes of 1,000 to 3,000 feet. In habitat the trees are magnificent, gnarled and moss-covered. In bloom, the trees can be seen from long distances since the white flowers act like a beacon in the forest. The large flowers are also mildly fragrant and have been commonly used as parents for the numerous hybrids available.

TURK'S CAP

Malvaviscus penduliflorus

Aloalo pahupahu

This tall shrub, a native of Mexico, is grown for its pendulous, bright-scarlet flowers that resemble hibiscus blooms that never open fully. The flowers were not often used in leis until the Micronesian style of lei making became popular a few years ago. The most common use of the plant is as an ornamental hedge or trimmed shrub in landscapes.

RED/CHINESE HIBISCUS

Hibiscus rosa-sinensis

This is the most popular hibiscus for hedges in Hawai'i. It is of Asian origin and will grow to 20 feet tall if not pruned back. Outside of Hawai'i this hibiscus is used for dyes, medicines, and fiber. The 4-inch flowers appear throughout the year and may be orange, magenta, yellow, and variegated, as well as red.

Anthuriums

When one first encounters an anthurium flower arrangement, it is easy to mistake the showy heart-shaped spathes and protruding flower spikes as being made of plastic. Anthurium leaves and blooms are very long-lasting and tough, which are great attributes to a florist. There are a large number of anthurium species which are in the family Araceae that includes the genus Philodendron and Monstera. Very few of the 500+ anthurium species have the showy spathes of Anthurium andreanum, which is also known as capotillo colorado, or the little red cape. The true flowers are tiny structures arranged spirally on the tail-like spadix. If pollination was successful, the spadix swells and berries develop.

ANTHURIUM

Anthurium andreanum

This species, the basis of a large cut-flower trade, is said to have been introduced by S. M. Damon in 1889, making it the first anthurium to reach Hawai'i. Today there are many hybrids and varieties that range from the common red to the white and green "obake" forms. The plants are also valued for their cut foliage, which is very long-lasting in arrangements.

Heliconias

❁ Heliconias are related to bananas but are in their own family, the Heliconiaceae. They have similar foliar appearances and range from 3 feet to over 20 feet tall, with both clumping and running growth habits. The true flowers are tubular and small and are secondary in appearance to the showy bracts. The inflorescences can be divided into erect and pendant forms. Each vegetative stalk grows, blooms, and then dies slowly as new stalks arise out of the underground rhizomes. Interestingly, the New World forms are primarily pollinated by hummingbirds, and the Old World forms are bat-pollinated. All are characterized by their striking form and wide range of colors. They are long-lasting and dramatic in tropical floral arrangements and are a mainstay in the Hawai'i cut-flower industry.

RAINBOW HELICONIA

Heliconia wagneriana

This popular heliconia blooms from January to September and is native from Belize to Colombia. It is commonly grown in landscapes because of its erect, undulate foliage and nonaggressive clumping habit.

BIRD OF PARADISE

Strelitzia reginae

The bird of paradise is a South African plant that is so commonly used in Hawai'i that many visitors consider it to be an island native. The inflorescences arise out of a clump of tough blue-green foliage. Orange sepals and blue, arrowhead-shaped flowers arise out of a green and red bract that looks like a bird's head. These colorful structures last two weeks as a cut flower and longer if left on the plant.

PARROT'S BEAK

Heliconia psitticorum

This is the most common heliconia, both in the landscape and as cut flowers. They bloom all year, with the inflorescences borne terminally on thin stalks that grow from rapidly running rhizomes. There are many varieties of this heliconia, with bract colors that range from green to yellow to orange and reds of various shades.

(left and right)

GIANT HELICONIA

Heliconia caribaea

Heliconia caribaea hails from the West Indies and grows to 20 feet tall with massive stalks. There are many forms and colors of this heliconia, which is popularly used in large arrangements. The bracts may be burgundy to pure red, or deep gold to whitish-yellow, or clear green with some varieties tinted or splotched with red or green.

RED COLLINSIANA

Heliconia collinsiana

This Central American heliconia has an attractive pendant inflorescence with red bracts and bright-yellow tubular flowers that bloom all year. A significant characteristic of the 15-foot plant is the copious coating of wax that gives the foliage and stems a whitish appearance.

Flowers Native to Hawai'i

The classification of Heritage plants is used here to include native endemic and indigenous flora, as well as early introduced species that have a long history with Hawai'i. Before Hawai'i became populated, it was a barren land that was slowly colonized by pioneer plants. These naturally dispersed species changed until 85 percent evolved into a unique species found nowhere else, known as endemics. The story of the colonization and evolution of the endemic flora of Hawai'i is a source of endless fascination about biodiversity to the inquisitive. Those species that arrived naturally but remain largely unchanged from types found outside of Hawai'i are known as indigenous species. The arrival of Polynesians began the introductions of plants by man. Since Captain Cook, Western introductions started the second wave of plant importations that continue to this day. The visitor wishing to see endemic species has an increasingly difficult task, since the recent history of Hawai'i is one of drastic habitat change. As a result, many endemic species are extinct or endangered. To the casual observer this sad state of affairs is easily overlooked, since Hawai'i is still lushly vegetated by exotic and alien species.

NATIVE WHITE HIBISCUS

Hibiscus arnottianus

Kokia keokeo

Hibiscus arnottianus is a medium tree of the rain forest areas in its native habitat. It adapts to the life of the dry lowlands but never achieves the majestic moss-draped and sinuous form of the wild plants. The large white blossoms are nicely complemented by the dark foliage and subtle fragrance.

MOUNTAIN APPLE

Eugenia malaccensis

This tall Indo-Malaysian native grows to 50 feet in wet forests up to 1,800 feet in elevation. The tufted red flowers bloom close to the trunk and branches of the tree in March and April. The forest floor is usually carpeted cerise after the flowers fall. Beginning in June, the thin-skinned, red fruits ripen. The juicy white flesh that surrounds a hard, round seed is very refreshing and sweet.

‘OHI‘A LEHUA

Metrosideros polymorpha

Yellow: Lehua mamo; White: Lehua puakea or ‘Ohi‘a kea

‘Ohi‘a lehua

The *‘ohi‘a lehua* is the most common Hawai‘i native tree in many forests from 1,000 to 9,000 feet in elevation. The specific name polymorpha alludes to the extreme variability of this plant, which can be a small shrub or a giant tree 100 feet tall. They are often associated with the native tree ferns and many *‘ohi‘a* trees germinate on tree ferns, later strangling their hosts. It is said that picking a *lehua* flower on the way to the mountains will cause it to rain. The blossoms provide nectar for many native birds and the colors range from white, yellow and salmon to red. Metrosideros means "heart of iron," and the dense, dark heartwood is very hard and durable.

ILIMA

Sidax fallax

'Ilima

'Ilima is a very variable plant that ranges from a sprawling ground cover to erect shrubs 4 feet high. The flower colors also vary from reddish to pale yellow. The plant is famous as representing O'ahu and is most commonly known as a deep-orange lei flower. Although it is found growing naturally along many coastal areas, *'ilima* is often cultivated for the lei industry. The *'ilima* leis were once reserved for royalty and today are a prized gift to receive. The flowers are about 1 inch across and extremely thin so that hundreds, or even thousands, of blossoms are required for a lei. Many *'ilima* leis are given with multiple strands which adds to the beautiful impression. Great labor is involved in this lei that is physically short-lived, but forever memorable.

PUA-KENIKENI

Fagraea berteriana

Pua-kenikeni is native to the South Pacific and grows into a spreading tree over 15 feet tall. The branches are quadrangular, with blunt-tipped leaves and heady yellow-orange flowers. The flowers are popular in leis and the Hawaiian name means "ten-cent flower," which is how much each blossom used to cost. Coconut oil was also laced with blossoms to infuse the flowers' perfume.

KOU

Cordia subcordata

Kou is a coastal tree about 30 feet tall that was introduced from Polynesia to Hawai'i by the early native immigrants. It was important to the Hawaiians as a source of wood for cups and bowls, highly prized for the beautiful grain. It is a popular shade tree and the orange, scentless flowers bloom in clusters throughout the year.

TI PLANTS

Cordyline fruticosa

Ki

Cordyline plants are originally from tropical Asia and the Pacific. Of extreme importance to the Hawaiians, ti was an early introduction with the migration of the native people. The plant is especially valued for its utility, and the leaves have been used for food wrapping, hula skirts, thatch, rope, sandals, raincoats, as well as religious ceremonies. The slender stalk topped by a rosette of leaves is believed to have served as the model for the royal standards, or *kahili,* that were beautifully crafted out of feathers. The flower stalk bears many tiny lilac-tinted blossoms followed by red or white berries. The common green ti rarely forms fruit but propagates readily from stem cuttings. There are now many varieties of ti with a wide range of leaf forms and color that make them especially valuable as landscape plants. A perimeter of ti around a house is said to ward off evil and harbor good luck.

Orchids

❁ Orchids are a very large, cosmopolitan family of plants, with 1,000 genera and 15,000 to 20,000 species, of which only three species are endemic to Hawai'i. Orchids have a long history of horticultural popularity, and Hawai'i has developed an important industry marketing the flowers and plants of many types. Orchids are terrestrial or epiphytic perennial herbs that are especially abundant in rain forest habitats. They are best known for the beauty of their flowers, which in many cases form fountains of color to please the viewer. Surprisingly, vanilla comes from the seed capsule of a vining orchid.

MOTH ORCHIDS

Phalaenopsis species

Moth orchids are an Asiatic genus that prefer humid, wet conditions and low light levels. Phalaenopsis plants have very short stems and seem to be stemless, with leathery leaves from 6 inches to 2 feet. Most popular are the large white varieties with flowers arranged in parallel rows on the floral stalk. These have long blooming seasons from winter to spring. Other types of Phalaenopsis may have small-to-large flowers in colors ranging from purple to yellow to pink and variously spotted or striped.

DENDROBIUM ORCHIDS

Dendrobium species

There are over 1,500 species of dendrobium orchids with countless hybrids and horticultural forms. The genus ranges from India to New Zealand and encompasses snow-covered habitat to the warm tropics. One reason for their popularity and spread is their ease of cultivation, long flowering season, and beautiful blossoms that are often arranged in large sprays. Coming from such a large group, there are miniatures to large cane-like forms 5 feet tall (before blooming). Dendrobium cultivation for the cut-flower trade is an important crop in Hawai'i. Most commonly grown are the tough hybrids whose blossoms are usually purple or white. Many local people have orchid collections and the hobbyist is likely to possess the odd and unusual species.

VANDA ORCHIDS

Vanda teres X V. hookeriana

Vanda hybrids like Vanda "Miss Joaquim," are grown as a major horticultural crop in Hawai'i for cut flowers and leis. Less important is the cultivation of about 70 other species, plus many hybrids for corsages, potted plants, and landscaping. Vandas are native from India to Indonesia and flower color ranges from lavender to purple to orange and various degrees of mottling. The blossoms may be 1 inch to 4 inches in diameter, arranged in few to many flowered inflorescences.

CATTLEYA ORCHIDS

Cattleya species

This group of orchids is originally from Brazil and, like many other orchids, is very popular for hybridizing. It is now easier to find hybrids than the true species and Cattleya is often used in bigeneric or trigeneric crosses. Their ease of culture and large, beautiful blooms make Cattleyas one of the most desirable orchids to grow. Cattleyas are grown for cut flowers, potted plants, and as accents in landscaping.

Bromeliads

❁ Bromeliads are a diverse group of New World plants that can be found growing in almost every habitat from the deserts to the rain forests. Some are on the ground growing as terrestrials or clinging to trees as epiphytes. Many bromeliads can be described as stiff rosettes of leaves, but Spanish moss is a form not often thought of as belonging to the family. Large numbers of species form tanks of water with their leaves and these types form little ecosystems that support a variety of life, including insects, frogs, and crabs. During dry periods, these tanks are often used as reservoirs to quench the thirst of passing monkeys and birds. Their flowers range from single, insignificant blooms to the massive inflorescences of Puya raimondi that rise 30 feet over the 10-foot plant. The most famous bromeliad is the pineapple, which has a long history of cultivation as a food crop.

Aechmea chantinii

There are many types of ornamental bromeliads. Some are grown for their attractive foliage, while others are known for their colorful inflorescences. Aechmea chantinii is but one form that has both beautiful leaves and a wonderful bloom. In its native habitat it is found as an epiphyte, growing in trees.

PINEAPPLE

Ananas comosus

Hala kahiki

Pineapples are a major agricultural crop in Hawai'i and large fields of the stiff, grey-ish rosettes are a common sight. Two years after planting, the fruits, which emerge above the leaves, can be harvested and eaten fresh or processed and canned.

INDEX